YOUR KNOWLEDGE HAS VALUE

- We will publish your bachelor's and master's thesis, essays and papers

- Your own eBook and book - sold worldwide in all relevant shops

- Earn money with each sale

Upload your text at www.GRIN.com
and publish for free

Nicotine Extraction from Tobacco Waste

Vani Gandham

Malyala Sai Krishna

Gond Parashuram

Gunja Anil

Reshma Lakra

Zakir Hussain

Thomas Lourdu Madanu

GRIN ☺

Bibliographic information published by the German National Library:

The German National Library lists this publication in the National Bibliography; detailed bibliographic data are available on the Internet at http://dnb.dnb.de.

ISBN: 9783389124123
This book is also available as an ebook.

© GRIN Publishing GmbH
Trappentreustraße 1
80339 München

Print and binding: Books on Demand GmbH, Norderstedt, Germany
Printed on acid-free paper from responsible sources.

The present work has been carefully prepared. Nevertheless, authors and publishers do not incur liability for the correctness of information, notes, links and advice as well as any printing errors.

GRIN web shop: https://www.grin.com/document/1575115

NICOTINE EXTRACTION FROM TOBACCO WASTE

Malyala Sai Krishna, Gond Parashuram, Gunja Anil, Vani Gandham, Reshma Lakra, Zakir Hussain, Thomas Lourdu Madanu

Department of Chemical Technology and Organic Chemistry, Loyola Academy, Secunderabad-500010, India

Contents

Abstract

Tobacco processing industries generate a large volume of tobacco waste as a by-product during the production of tobacco products. These wastes are typically discarded through methods like incineration or burial. However, this discarded material still contains significant economic value that is often overlooked. One of the most valuable components of tobacco waste is its high nicotine content. Nicotine is a key raw material for a variety of products, including those in the pharmaceutical and personal care industries, particularly to produce hygienic products such as nicotine patches or e-cigarette liquids. Therefore, finding ways to extract and utilize the nicotine from tobacco waste could have both economic and environmental benefits. The extraction of nicotine from tobacco waste is an important issue, as it presents an opportunity to create a valuable resource for various industries, while also mitigating the environmental harm caused by the burning of tobacco waste. Incineration and burial contribute to air and soil pollution, which can have long-lasting negative impacts on the environment. By extracting nicotine from these wastes, industries can repurpose what would otherwise be discarded material, reducing waste and contributing to more sustainable practices. This experiment focuses on a simplified approach to nicotine extraction from tobacco waste using an alkaline extraction method. In this process, sodium hydroxide (NaOH) solution is used to break down the plant material and release nicotine. Following this, diethyl ether is employed for solvent separation to isolate the nicotine from the extract. To confirm the presence of nicotine, picric acid and methanol are added to the solution, inducing the formation of a picrate precipitate. The appearance of a colorless precipitate after cooling the mixture serves as a positive indicator **of** the presence of nicotine in the extract. This approach provides a relatively straightforward and effective method for extracting nicotine from tobacco waste, offering a potential pathway for repurposing this by-product into valuable resources while reducing environmental impact.

1. Introduction

The extraction of nicotine from tobacco waste presents a sustainable avenue for resource recovery and potential pharmaceutical applications. Tobacco waste, often discarded, contains significant nicotine concentrations, offering an alternative to traditional cultivation. This introduction explores the development of efficient and environmentally conscious extraction methodologies. Focusing on solvent extraction and other advanced techniques, this research aims to optimize nicotine yield while minimizing environmental impact. The extracted nicotine can be utilized in nicotine replacement therapies and other pharmaceutical products, thus transforming waste into a valuable resource [1].

1.1. Tobacco

Tobacco leaves, the heart of the *Nicotiana tabacum* plant, are the source of an array of chemical compounds, most notably nicotine. These leaves, varying in size, shape, and texture depending on the cultivar and growing conditions, are the primary raw material for a diverse range of products, from cigarettes and cigars to smokeless tobacco and, increasingly, pharmaceutical applications. The chemical composition of tobacco leaves is intricate, encompassing alkaloids, terpenes, polyphenols, and various other organic compounds. Nicotine, a potent neurotoxin, constitutes a significant portion of the alkaloid content, typically ranging from 0.3% to 5% of the dry weight. However, the exact concentration is influenced by factors such as the variety of tobacco, the leaf's position on the stalk, and the curing process. Leaves located higher on the stalk generally exhibit higher nicotine levels. Beyond nicotine, tobacco leaves contain a plethora of other alkaloids, including nicotine, anatabine, and anabases, albeit in smaller quantities. These minor alkaloids, while present at lower concentrations, play a role in the overall flavor and aroma profile of tobacco products. Furthermore, they are of scientific interest due to their potential conversion into carcinogenic tobacco-specific nitrosamines (TSNAs). Polyphenols, such as chlorogenic acid and rutin, are known for their potential health benefits, although these benefits are often overshadowed by the detrimental effects of nicotine and other harmful compounds in tobacco smoke [2].

4

Fig 1: Dry Tobacco Leaves [2]

The curing process, a crucial step in preparing tobacco leaves for use, significantly impacts their chemical composition. Curing methods, including air-curing, flue-curing, sun-curing, and fire-curing, influence the levels of sugars, nicotine, and other compounds. Flue-curing, for instance, reduces sugar content while maintaining nicotine levels, resulting in a characteristic flavor profile.

The positional variation of chemical constituents within the leaves along the stalk is a significant aspect of tobacco chemistry. Upper leaves have a higher concentration of nicotine, while lower leaves have a higher concentration of sugars. This variance in chemical composition is critical for determining the suitability of leaves for specific applications.

In recent years, research has explored the potential of tobacco leaves beyond traditional smoking products. Studies have investigated the extraction of valuable compounds for pharmaceutical and industrial applications. Nicotine, for example, is being explored for its potential therapeutic uses in smoking cessation and neurodegenerative diseases. Additionally, the leaves are also being investigated for the extraction of other potentially useful compounds.

However, the overwhelming association of tobacco leaves with harmful smoking products necessitates careful consideration of the ethical and public health implications of any research or application. The addictive nature of nicotine and the carcinogenic potential of other tobacco constituents remain significant concerns.

1.1.1. Tobacco Waste in Worldwide and India

Global Tobacco Market

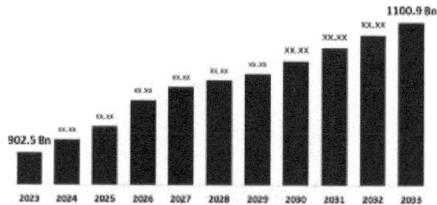

Fig 2(a): The Global Tobacco Market Over [3]

Tobacco, a crop with a complex history and significant economic impact, is cultivated worldwide. Its production, however, is concentrated in a few key nations, shaping the global tobacco market. China stands as the undisputed leader, followed by Brazil, India, and the United States. These countries, along with others in Africa and Southeast Asia, contribute substantially to the global supply of raw tobacco.

The demand for tobacco, though facing increasing regulatory pressures and health awareness campaigns, remains significant, driving a complex network of cultivation, processing, and trade. The global tobacco trade is characterized by the movement of raw tobacco from producing nations to manufacturing hubs, primarily located in developed and developing countries. This trade is influenced by factors such as quality, price, and regulatory frameworks, leading to a dynamic and often volatile market. Global organizations and agreements, such as the Framework Convention on Tobacco Control (FCTC), play a role in shaping the regulatory landscape, aiming to curb tobacco consumption and its associated health risks [3].

1.1.2 India's Prominent Role

India's position as the world's second-largest producer of tobacco is a significant aspect of the global agricultural and economic landscape.

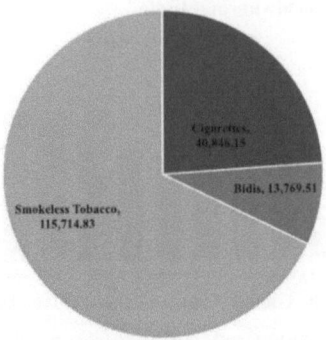

Fig2(b): Annual Waste of Tobacco Product [4]

This prominence is not merely a statistical fact but a result of a confluence of favorable agroclimatic conditions, deeply entrenched agricultural practices, and a substantial domestic and international market. The country's contribution extends beyond production volumes, encompassing its role as a major exporter of unmanufactured tobacco, a vital source of foreign exchange, and a provider of livelihoods for millions.

Within this global context, India occupies a position of considerable significance. It is the world's second-largest producer of tobacco, a testament to its favorable agroclimatic conditions and established agricultural practices. India is also a major exporter of unmanufactured tobacco, contributing significantly to the global supply chain. This export-oriented approach is crucial for India's economy, generating foreign exchange and supporting numerous livelihoods [4].

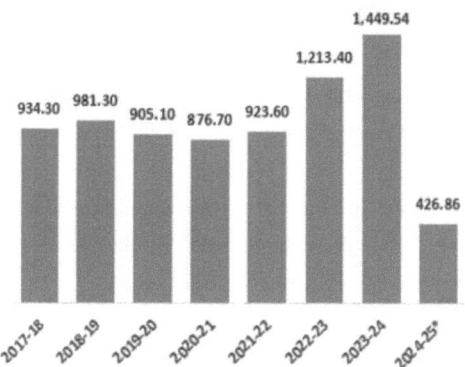

Fig 3: Tobacco Exports over time [5]

India's tobacco production is diverse, encompassing various types of tobacco tailored to specific market demands. Flue-cured Virginia (FCV) tobacco, a light-colored leaf used primarily in cigarette manufacturing, is a significant component of India's production. India also produces substantial quantities of bidi tobacco, a dark, strong-flavored leaf used in the production of bidis, a traditional Indian smoking product. Additionally, chewing tobacco, another prevalent form of consumption in India, contributes to the overall production volume [5].

1.1.3 Geographical Distribution and Agricultural Practices

The cultivation of tobacco in India is concentrated in specific regions, primarily in the southern states of Andhra Pradesh, Karnataka, and Telangana. These states possess the ideal soil and climatic conditions for tobacco growth, fostering a robust agricultural sector dedicated to this crop. The Tobacco Board of India, a statutory body, plays a crucial role in regulating and promoting the tobacco industry, ensuring quality control, and supporting farmers.

Indian tobacco farming practices vary depending on the type of tobacco being cultivated. FCV tobacco, for instance, requires meticulous management, including controlled curing processes to achieve the desired leaf characteristics. Bidi tobacco, on the other hand, is often grown in less intensive systems, relying on traditional farming methods. The agricultural practices employed in tobacco cultivation have significant implications for soil health, water resources, and environmental sustainability [6].

8

1.1.4 Economic and Social Impact

The tobacco industry in India has a profound economic and social impact. It **employs** millions of people, particularly in rural areas where tobacco cultivation is concentrated. These jobs range from farming and processing to trading and manufacturing. The industry also contributes significantly to the national economy through revenue generation and foreign exchange earnings. However, the tobacco industry also faces increasing scrutiny due to the well-documented health risks associated with tobacco consumption. Public health campaigns and regulatory measures aim to reduce tobacco usage, posing challenges to the industry's long-term sustainability. The Indian government is navigating a complex balance between supporting the livelihoods of tobacco farmers and addressing the public health implications of tobacco consumption.

1.1.5 Challenges and Future Directions

The Indian tobacco industry faces several challenges, including fluctuating market prices, increasing competition from other tobacco-producing nations, and evolving regulatory landscapes. The industry is also grappling with the need to adopt sustainable agricultural practices to minimize its environmental footprint. Looking ahead, the Indian tobacco industry must adapt to these challenges to ensure its long-term viability. This may involve diversifying into alternative crops, adopting sustainable farming practices, and investing in research and development to enhance productivity and quality. The industry must also engage with public health initiatives to address the concerns surrounding tobacco consumption. **Finally**, India's tobacco production is a complex and multifaceted sector, playing a significant role in both the national economy and the global tobacco market. While the industry faces challenges, it also possesses the potential to adapt and evolve, contributing to sustainable development and economic growth.

Globally, tobacco cultivation and processing generate substantial quantities of waste, posing environmental challenges. In India, a major tobacco producer, the management of tobacco waste is of particular concern. However, recent research indicates that this waste stream can be repurposed for biochar production, offering a sustainable solution for both waste management and soil disease control. Utilizing tobacco waste as a biochar source aligns with the principles of circular economy and provides an environmentally friendly alternative to traditional waste disposal methods. This study investigates the utilization of biochar derived from agricultural waste, including potential tobacco waste, for the control of black shank disease [7].

1.2 Nicotine

Nicotine, a naturally occurring alkaloid extracted from *Nicotiana tabacum*, commands significant attention across diverse scientific disciplines. Its potent interaction with nicotinic acetylcholine receptors in the central and peripheral nervous systems drives its complex pharmacological profile, characterized by both stimulatory and depressant effects. This duality underpins nicotine's addictive nature, a critical consideration in public health, while also sparking interest in its potential therapeutic applications [8].

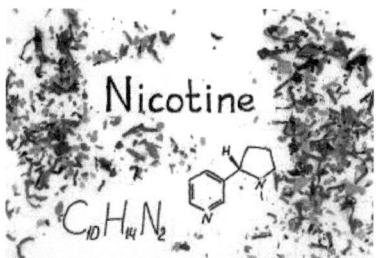

Fig 4: Nicotine from tobacco waste [8]

Research explores nicotine's influence on cognitive function, neurodegenerative diseases like Parkinson's and Alzheimer's, and mental health disorders. Though historically associated with tobacco dependence and its severe health consequences, the compound's isolated effects reveal a spectrum of physiological responses. Understanding the intricate mechanisms of nicotine's action is crucial for developing effective smoking cessation strategies and exploring its potential in novel therapeutic interventions. This introduction will delve into the multifaceted aspects of nicotine, examining its chemical properties, physiological impacts, and diverse applications, while acknowledging the inherent ethical and health-related complexities.

1.2.1 Chemical Formula and Structure

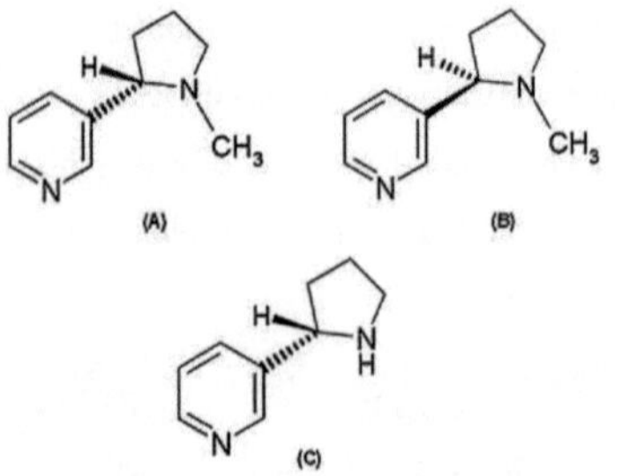

Fig 5: Structure of Nicotine [10]

Nicotine, the predominant alkaloid in Nicotiana tabacum, has been the subject of extensive research due to its central role in tobacco dependence. While nicotine constitutes approximately 95% of the total alkaloids, minor alkaloids, including nor-nicotine, anatabine, and anatabine, are also present, each representing 0.3 to 3% of the alkaloid composition. Notably, nicotine is chiral, with N tabacum primarily synthesizing the enantiomer, a characteristic consistent with the enantioselectivity observed in many biological systems. The potential therapeutic application of both nicotine enantiomers in smoking cessation has been recognized [9]. However, during metabolic processes, these enantiomers undergo structural transformations, yielding carcinogenic metabolites. Consequently, the pharmacological effects of the nicotine enantiomer, which are posited to differ significantly from those of the dominant enantiomer, remain relatively unexplored. This knowledge gap is largely attributed to the absence of efficient purification methodologies for the antipode. Conventional purification techniques are time-consuming and yield insufficient quantities for comprehensive pharmacological studies. Furthermore, the lack of a viable asymmetric synthetic route has further impeded research into this enantiomer [10]. This limited investigation is likely due to the minimal human exposure and intake of nicotine. This study addresses these limitations by introducing a novel and efficient methodology for the extraction and purification of bioactive compounds from waste tobacco stems. This approach aims to facilitate the isolation of nicotine, enabling a more thorough investigation of its pharmacological properties. Moreover, it underscores the potential for the sustainable utilization of agricultural by-products and contributes to the advancement of green chemistry principles. The extract exhibiting the highest nicotine (NCT) concentration, obtained from the upper tobacco leaf fraction via acid-

11

base extraction, was selected for incorporation into a fast-dissolving film formulation. The optimized film was subsequently characterized for its physicochemical and mechanical properties, in vitro disintegration profile, and drug release kinetics.[11] Characterization revealed that the extract derived from the upper tobacco leaf fraction, utilizing the aforementioned extraction methodology, yielded the highest NCT content. The resulting NCT fast-dissolving film, formulated with this extract as the active pharmaceutical ingredient and hydroxypropyl methylcellulose (HPMC) E15 as the film-forming polymer, presented as a homogeneous, translucent film with a light brown hue. The incorporation of NCT significantly modulated the film's physicochemical and mechanical attributes, as evidenced by alterations in weight, disintegration time, tensile strength, percentage elongation at break, and Young's modulus. The in vitro drug release profile of the NCT fast-dissolving film demonstrated a rapid initial release, with 80% of the drug released within three minutes. Furthermore, the drug release kinetics were best described by the Higuchi matrix model, indicating a diffusion-controlled release mechanism. These findings suggest that the developed NCT fast-dissolving films possess desirable characteristics for potential clinical application, offering rapid drug release and favorable physicochemical properties. The positional variation of chemical composition within tobacco leaves along the stalk provides critical insights into their suitability for specific applications [12]. Nicotine (NCT), a prominent alkaloid constituent of Nicotiana tabacum leaves, exhibits solubility in a range of solvents, including alcohols, chloroform, ethers, petroleum ether, kerosene, and water. This solubility profile facilitates the extraction of NCT from tobacco leaves via solvent extraction methodologies. NCT finds diverse applications across the fine chemical, pharmaceutical, and agricultural industries, as well as within the tobacco industry itself, where it functions as a key cigarette additive. In pharmaceutical applications, NCT is employed in smoking cessation therapies to alleviate withdrawal symptoms. However, oral administration of nicotine replacement therapy (NRT) is often limited by significant first-pass metabolism, leading to rapid NCT degradation. To address these limitations, various NCT dosage forms have been developed, including transdermal patches, mouth sprays, lozenges, chewing gum, and oral films. Each delivery system presents distinct advantages and disadvantages. Transdermal patches, for instance, offer a slow, sustained-release profile, which contrasts with the rapid bolus delivery associated with cigarette smoking.

However, they are frequently associated with localized skin reactions at the application site,

12

necessitating daily rotational application. While both skin and buccal mucosa exhibit permeability, oral mucosal regions, particularly the buccal mucosa, composed of non-keratinized cells, demonstrate significantly higher permeability compared to skin. Mouth sprays can thus enhance NCT absorption. Consequently, a detailed understanding of the chemical composition of tobacco leaves, particularly their positional distribution on the stalk, is crucial for optimizing extraction and formulation strategies tailored to specific applications. Nicotinic acid also referred to as niacin or vitamin B3, is employed in the therapeutic management of hyperlipoproteinemia classified as types II, III, IV, and V, due to its capacity to reduce circulating cholesterol and triglyceride levels. This hypolipidemic effect is attributed, in part, to nicotinic acid's ability to enhance vascular endothelial cell lipoprotein lipase activity, thereby augmenting very low-density lipoprotein (VLDL) clearance. Furthermore, nicotinic acid diminishes the circulating levels of free fatty acids mobilized from adipose tissue. The therapeutic dosage of nicotinic acid required for cholesterol and triglyceride reduction significantly exceeds that used for vitamin supplementation. Niacin is currently available in immediate-release, sustained-release, and extended-release formulations. To mitigate the incidence of adverse effects, nicotinic acid therapy is typically initiated at a low dose and gradually titrated upwards.

The primary adverse effect associated with nicotinic acid is flushing a prostaglandin-mediated response. Other reported adverse effects include pruritus, exacerbation of peptic ulcers, hyperuricemia, and impaired glucose tolerance. Notably, hepatic necrosis has been associated with sustained-release formulations of nicotinic acid. It is crucial to recognize that doses of extended-release nicotinic acid are not interchangeable with those of immediate-release preparation. Nicotinic acid and various nicotinamide derivatives serve as precursors for the synthesis of nicotinamide adenine dinucleotide, a crucial coenzyme synthesized in mitochondria and essential for oxidative energy production in numerous metabolic reactions.

1.3 Methods Available to Extract Nicotine Acid

1.3.1 Solvent Extraction:

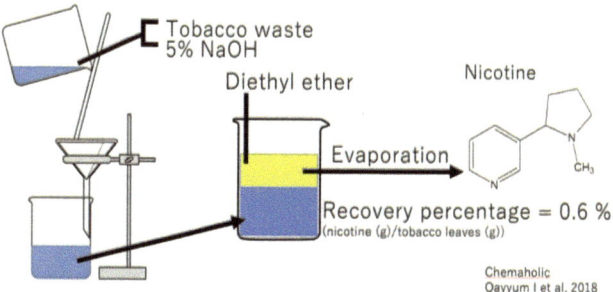

Fig 6: Solvent Extraction [13]

Nicotine extraction typically employs solvent extraction, leveraging its solubility in organic solvents. Tobacco waste, often pre-treated with an alkaline solution to liberate free-base nicotine, is mixed with a suitable solvent like dichloromethane or ether. Nicotine, being more soluble in these solvents than in the aqueous phase, partitions into the organic layer. This organic layer, containing the extracted nicotine, is then separated. Subsequent steps, such as evaporation or distillation, are isolating the nicotine from the solvent. The choice of solvent and pH control are critical for efficient extraction [13].

The efficiency of nicotine extraction is heavily influenced by several factors, including the choice of solvent, pH control, temperature, and contact time. Selecting a solvent with high selectivity for nicotine and low solubility in water maximizes extraction efficiency. Maintaining an optimal pH, usually in the alkaline range, ensures that nicotine remains in its free-base form. Temperature affects the solubility and diffusion rate of nicotine, while contact time allows for adequate partitioning between the phases. Careful control of these parameters is essential for obtaining a high yield of pure nicotine [14].

1.3.2 Acid-Base Extraction:

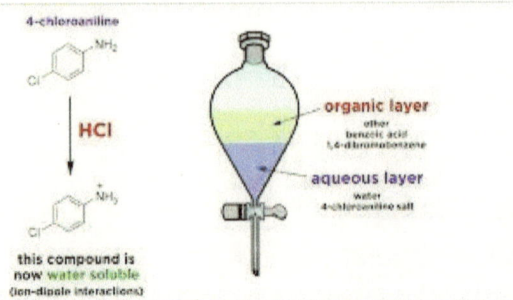

Fig 7: Acid-Base Extraction [15-16]

The acid-base extraction method stands as a cornerstone for isolating nicotine from tobacco leaves, leveraging the compound's alkaloid nature to achieve a higher degree of purity compared to simpler extraction techniques like maceration. This method's effectiveness stems from its ability to manipulate nicotine's solubility by altering the pH of the extraction medium, effectively shifting the equilibrium between its free base and salt forms. The initial step involves treating the powdered tobacco leaves with a basic solution, typically sodium carbonate. This process aims to liberate nicotine from its salt form, which is prevalent within the plant matrix, into its free base form. Nicotine, in its free base state, exhibits significantly higher solubility in organic solvents compared to water, a crucial characteristic exploited in subsequent extraction steps. Following the basification, an organic solvent, such as chloroform, is introduced to extract the free base of nicotine from the aqueous solution [15]. This liquid-liquid extraction technique relies on the principle of partitioning, where nicotine preferentially dissolves in the organic solvent, leaving behind water-soluble impurities. Multiple extractions with fresh organic solvent ensure maximum recovery of nicotine from the aqueous phase. The combined organic extracts, now enriched with nicotine, are then subjected to concentration, commonly via rotary evaporation, to remove the volatile solvent and obtain a concentrated nicotine extract. This acid-base extraction method not only yields a relatively pure nicotine extract but also offers flexibility in preparing nicotine for diverse applications. The ability to control nicotine's solubility through pH manipulation, coupled with the efficiency of liquid-liquid extraction, makes this method a preferred choice for isolating nicotine from tobacco leaves [16].

15

1.3.4 Supercritical Fluid Extraction (SFE):

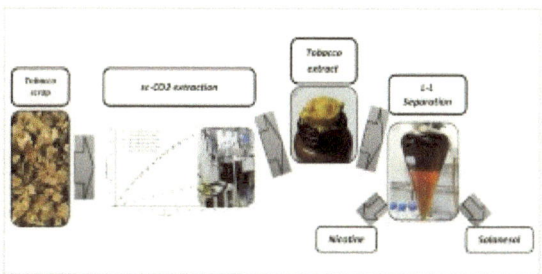

Fig 8: Supercritical Fluid Extraction (SFE) [17-18]

Supercritical fluid extraction (SFE) presents a modern and efficient approach to nicotine extraction, diverging significantly from traditional solvent-based methods. At its core, SFE utilizes supercritical fluids, most commonly carbon dioxide (CO_2), as the extraction solvent. A supercritical fluid exists in a state where it exhibits properties of both a gas and a liquid, offering unique advantages for extraction processes. This state is achieved by manipulating temperature and pressure beyond the fluid's critical point.[17] The primary advantage of SFE lies in its enhanced extraction efficiency and selectivity. By adjusting temperature and pressure, the solvent's density and solvation power can be precisely controlled, allowing for targeted extraction of nicotine while minimizing the co-extraction of unwanted compounds. This level of control is particularly valuable for obtaining high-purity nicotine extracts. Furthermore, SFE can be considered an environmentally friendly alternative to traditional solvent extraction. Supercritical CO_2 is non-toxic, non-flammable, and readily available. After extraction, the Odett et al., 1984 can be easily removed by reducing pressure, leaving behind a solvent-free extract. This eliminates the need for extensive solvent removal and purification steps, reducing waste and environmental impact. In addition to its environmental benefits, SFE offers several operational advantages. The method's ability to achieve high extraction rates and yields translates to increased productivity. The adjustable parameters of SFE also allow for optimization based on the specific characteristics of the plant material and the desired nicotine purity. This method can be used for the extraction of nicotine from tobacco leaves, and also from tobacco waste, which is very useful for the environment. Overall, SFE represents a sophisticated and sustainable approach to nicotine extraction, offering a compelling alternative to conventional extraction methodologies [18].

1.4 Nicotine Percentage Available in Tobacco Waste:

Nicotine is a naturally occurring alkaloid in the tobacco plant. The concentration of nicotine within the plant itself can vary due to factors like the specific tobacco variety, growing conditions (soil, climate), and harvesting methods. Generally, nicotine constitutes roughly 0.6–3.0% of the dry weight of tobacco.

Nicotine **is** a potent water-soluble alkaloid extracted from the leaves of Nicotiana tabacum presents a significant public health concern due to its profound neurotoxic effects and high addictive potential. The processing of tobacco leaves, through drying and shredding, facilitates its consumption in various forms, including snuff, chewing tobacco, cigarettes, cigars, and pipe tobacco, each contributing to widespread nicotine exposure. Beyond recreational use, nicotine finds applications in smoking cessation therapies and as an insecticide, underscoring its diverse pharmacological and toxicological properties. The concentration of nicotine within tobacco leaves ranges from 0.5% to 8.0% (w/w), while cigarettes typically contain 10-20 mg/g of tobacco, highlighting the substantial quantities of this alkaloid delivered during consumption [19].

The nicotine percentage in tobacco waste within India exhibits considerable variability, generally ranging from approximately 1% to 3%, though specific instances may deviate from this range. This variation is attributed to several factors, including the diverse types of tobacco cultivated, such as Virginia, Natu, and Rustica, and the specific plant parts utilized, whether leaves or stems. Furthermore, differing processing methods and manufacturing stages contribute to the fluctuating nicotine concentrations. India's extensive tobacco industry, encompassing cigarettes, bidis, chewing tobacco, and hookah tobacco, generates a wide array of waste products. Notably, bidi manufacturing and the production of smokeless tobacco contribute significantly to the overall volume of tobacco waste. Despite the inherent variability in nicotine content, efforts are made to extract nicotine from this waste for applications such as pesticide production, specifically nicotine sulfate, though stringent regulations govern its use. It is crucial to acknowledge the health risks associated with nicotine and tobacco waste, underscoring the necessity for strict regulations and safety protocols in handling and processing these materials.

1.5 Applications of Nicotine

1.5.1 Nicotine Replacement Therapy (NRT):

Smoking Cessation:

This is the most established therapeutic application. NRT products (patches, gums, lozenges, sprays, inhalers) deliver controlled doses of nicotine to alleviate withdrawal symptoms, aiding in smoking cessation. This work could delve into the efficacy of different NRT forms, combination therapies, and factors influencing success rates.

Harm Reduction:

Exploring the concept of NRT as a harm reduction strategy for individuals unable or unwilling to quit smoking entirely.

1.5.2 Potential Therapeutic Applications Beyond Smoking Cessation:

Neurodegenerative Diseases:

Research explores nicotine's potential in mitigating symptoms of diseases like Parkinson's disease, Alzheimer's disease, and other forms of cognitive decline.

Thesis work could examine the mechanisms of action, clinical trial results, and the potential for nicotine-based therapies.

Mental Health Disorders:

Studies investigate nicotine's effects on conditions like schizophrenia and attention deficit hyperactivity disorder (ADHD).

This work could analyze the role of nicotinic acetylcholine receptors in these disorders and the potential for targeted therapies.

Pain Management:

Nicotine's analgesic properties are being studied for potential applications in chronic pain management.

1.5.3. Agricultural Applications:

Insecticide:

Historically, nicotine has been used as an insecticide.

Thesis work could explore the history of its use, its environmental impact, and the development of safer alternatives.

Addiction and Dependence:

Acknowledge the highly addictive nature of nicotine and the risks associated with its use.

Discuss the mechanisms of nicotine addiction and the challenges of dependence.

Adverse Effects:

Thoroughly address the potential adverse effects of nicotine, including cardiovascular risks, gastrointestinal issues, and other health concerns.

Ethical Considerations:

Discuss the ethical implications of nicotine use, particularly in vulnerable populations.

Current Research and Future Directions:

Provide an overview of current research trends and p

This outline presents a multifaceted view of nicotine's applications, extending beyond its notorious role in addiction. Nicotine Replacement Therapy (NRT) is well-established for smoking cessation, but research explores its potential in harm reduction for those unable to quit. Emerging applications include mitigating neurodegenerative diseases, managing mental health disorders, and chronic pain relief, though these remain largely experimental. Historically, nicotine's use as an insecticide has also been noted.

2. Literature Review

Kheawfu et al. research explores the development of a fast-dissolving oral film for nicotine (NCT) delivery as a potential alternative to traditional nicotine replacement therapies. The study focuses on optimizing NCT extraction from Burley tobacco leaves, comparing maceration and acid-base methods across different leaf stalk positions. Acid-base extraction from upper leaves yielded the highest NCT content, which was subsequently formulated into a thin film using HPMC E15. The resulting film exhibited desirable physical properties, including rapid in vitro disintegration and a swift 80% NCT release within three minutes, adhering to the Higuchi matrix model. The incorporation of NCT significantly influenced film characteristics, demonstrating the feasibility of creating a fast-dissolving film for clinical NCT administration. This work highlights the potential of tobacco-derived NCT in novel drug delivery systems, emphasizing the importance of extraction method optimization and formulation studies for effective therapeutic applications [20].

Ranjana Singh et al. study investigates nicotine extraction from various tobacco products, focusing on comparative analysis. Nicotine, a naturally occurring alkaloid, was extracted using a liquid-liquid extraction method with diethyl ether. The research aimed to quantify and compare nicotine levels in non-filtered tobacco (beedi/bidi), cigarettes, and chewing tobacco. The findings reveal that beedi/bidi contained the highest nicotine concentration, surpassing that of manufactured cigarettes. While chewing tobacco displayed lower nicotine levels per unit, its frequent consumption resulted in comparable daily nicotine intake to smoking. This study underscores the variability in nicotine content across different tobacco products, emphasizing the importance of considering consumption patterns when assessing nicotine exposure. The research contributes to the understanding of nicotine distribution in diverse tobacco forms, providing valuable data for public health considerations [21].

R. Megharaj et al. research addresses the environmental challenge posed by waste tobacco materials from cultivation and cigarette manufacturing, focusing on the efficient extraction of valuable compounds. A novel column-chromatographic extraction (CCE) method was developed to simultaneously recover nicotine and solanesol from tobacco leaf veins, followed by automated separation and simplified purification. Employing a petroleum ether/alkali ethanol solvent mixture, the CCE procedure achieved over 96% extraction efficiency for both compounds. Separation at pH 2.0 yielded a nicotine-rich ethanol-aqueous phase and a solanesol-containing ether phase. Further purification steps, including vacuum concentration and ether fractionation for nicotine, and silica gel chromatography for solanesol, resulted in high-purity products. Notably, the entire process was conducted at room temperature, with complete solvent recovery for reuse. This study presents a streamlined and cost-effective approach to valorize tobacco waste, mitigating environmental contamination while recovering valuable industrial compounds [22].

Ferguson S.G. et. al.'s research addresses the critical issue of nicotine extraction for therapeutic purposes, driven by the detrimental health impacts of tobacco smoke's complex chemical composition. Nicotine, a highly addictive compound, is targeted for extraction to facilitate smoking cessation through alternative delivery methods. This study employed a deep eutectic solvent, choline chloride/ethyl alcohol, to extract high-purity nicotine from cigarette tobacco. Infrared analysis confirmed the successful extraction of nicotine as a fatty substance. The investigation explored the influence of temperature and extraction time on nicotine yield, revealing optimal conditions at 60 minutes and 70°C, achieving a maximum extraction rate of 8.14%. Notably, the crude extract's purity was significantly enhanced from 65% to 98% through subsequent processing. This work contributes to the development of efficient nicotine extraction techniques, essential for creating effective smoking cessation aids and mitigating the adverse health consequences associated with traditional tobacco consumption [23].

Anyiam Ngozi Donald et. al. literature review focuses on *Nicotiana tabacum L.* (NTL) as the primary source of nicotine, tracing its origins and commercial significance. NTL belongs to the Solanaceae family, which also includes common food crops. The term "nicotine" derives from Jean Nicot de Ville Main, who introduced tobacco to France, highlighting its historical medicinal use. Tobacco's journey from Brazil to Europe underscores its widespread adoption. Today, *Nicotiana tabacum L.* is a globally significant commercial crop, with substantial cultivation and consumption, particularly in China. This overview emphasizes the plant's botanical classification, historical context, and current industrial relevance as a key source of nicotine.

Tobacco's transatlantic journey from Brazil to Europe catalyzed its rapid global spread, transforming *Nicotiana tabacum L.* into a commercially vital crop. Today, China leads in its cultivation and consumption, highlighting its enduring industrial significance. Botanically, it belongs to the Solanaceae family, known for producing nicotine, a potent alkaloid. Its historical context, intertwined with colonialism and trade, shaped its present-day ubiquity. The plant's adaptability and the addictive nature of nicotine have solidified its position in global agriculture and industry, despite increasing health concerns [24].

21

Arie Febrianto Mulyadi et al. study focuses on optimizing nicotine extraction from *Nicotiana tabacum L.*, exploring its potential beyond traditional cigarette production. Recognizing nicotine as a hazardous component but also a valuable agricultural pesticide, the research investigates solvent extraction using ether and petroleum ether. The study employs a response surface methodology with a central composite design to determine the optimal solvent combination for maximizing nicotine yield and minimizing extraction time. The findings reveal a significant impact of solvent ratios on both extraction time and yield. The predicted optimal solution, using 59.46 ml of ether and 30.12 ml of petroleum ether, resulted in an extraction time of 682.483 seconds and a nicotine yield of 4.80617%. This research provides valuable insights into efficient nicotine extraction, highlighting its potential for alternative applications and emphasizing the importance of optimized solvent ratios for maximizing yield and minimizing processing time [25].

Almeida R.N. et al. research investigate the extraction and characterization of nicotine from *Nicotiana tabacum L.*, specifically focusing on tobacco leaves extracted from Gold Live Classic Brand™ cigarettes. The study employs a liquid-liquid solvent extraction method using ether, preceded by dissolving the leaves in a sodium hydroxide (NaOH) solution. This approach aims to isolate nicotine, a key alkaloid, from the complex matrix of cigarette tobacco. The initial extraction yielded a low percentage of 0.6%, indicating a significant loss of product during the procedure. Potential causes for this low yield were identified, including the formation of emulsions and insufficient washing with ether to maximize nicotine recovery. To address this, the extraction process was repeated three times, highlighting the challenges associated with achieving efficient nicotine extraction from this particular cigarette brand. This iterative approach underscores the importance of optimizing extraction protocols to minimize product loss and improve yield.

To confirm the identity of the extracted compound, various physical properties were determined. The molecular weight (MW) was found to be 162.23 g/mol, the melting point (MP) at -79°C, and the boiling point (BP) at 246.8°C. Furthermore, the specific rotation ($[\alpha]D$) of nicotine was determined to be -168.5° at 293.15 K. These physical characteristics align with established data for nicotine, providing strong evidence for its successful extraction. Additionally, infrared (IR) spectroscopy was utilized to further verify the presence of nicotine. The IR spectra revealed distinct peaks corresponding to the bond frequencies of specific functional groups characteristic of nicotine. This spectral analysis corroborated the physical property data, confirming the successful isolation and identification of nicotine from cigarette tobacco. This study highlights the challenges and considerations involved in extracting nicotine from commercial cigarette products. The low initial yield emphasizes the need for optimized extraction techniques to improve efficiency and minimize product loss. Furthermore, this research adds to the existing knowledge of nicotine extraction, especially within the context of commercial cigarette tobacco, and underscores the importance of methodical extraction and characterization techniques. This research details the extraction of nicotine from Gold Live Classic Brand™ cigarettes using liquid-liquid extraction with ether and NaOH. A low initial yield (0.6%) necessitated repeated extractions, revealing challenges in maximizing nicotine recovery [26].

Maheshbabu et al. research delves into the intricate chemical composition of tobacco products, transcending the simplistic view of nicotine as the sole driver of addiction. It emphasizes the synergistic interplay between nicotine and other critical components, such as tar and various additives, to elucidate the complex mechanisms underlying tobacco dependence. The study recognizes that the addictive nature of tobacco is not solely attributable to nicotine but is significantly influenced by the combined effects of these chemical mixtures.

To achieve a comprehensive understanding, the study employs advanced analytical techniques, specifically gas chromatography-mass spectrometry (GC-MS) and high-performance liquid chromatography (HPLC). These methods are instrumental in profiling the diverse chemical constituents present in various tobacco samples. By utilizing these sophisticated techniques, the research aims to quantify and analyze the ratios of nicotine, tar, and other additives, revealing significant variations across different tobacco products.

The findings highlight that the ratios of these components are not uniform, but rather exhibit substantial variability. This variability is crucial as it underscores the potential for synergistic effects, where the combined presence of certain chemicals amplifies nicotine's addictive potential. For instance, certain additives may enhance nicotine absorption or alter its metabolism, thereby increasing its impact on the brain's reward pathways. Similarly, tar, a complex mixture of combustion byproducts, contributes to the overall toxicity and may also influence nicotine's addictive properties through various mechanisms.

This study aims to provide a detailed chemical profile of tobacco products, offering valuable insights into the dynamics of addiction. This comprehensive profile is essential for understanding the complex interactions between various chemical components and their combined impact on health. By quantifying and analyzing these ratios, the research contributes to a deeper understanding of the mechanisms driving tobacco dependence [27].

3. Material and Methods

3.1 Materials and Chemicals

Tobacco Waste: The raw material collected from BBM cigarette factory, Jeedimetla, Hyderabad.

Distilled Water: A polar solvent, used in water maceration.

Ethanol: A moderately polar solvent, used in ethanol maceration.

Sodium Carbonate ($Na_2 CO_3$): An alkaline solution, used in acid-base extraction.

Sodium Hydroxide (NaOH): A strong alkaline solution, used in acid-base extraction.

Diethyl Ether ((C_2H_5)$_2$O: An organic solvent, used in acid-base extraction.

All these chemicals were procured from Sigma Aldrich, Hyderabad.

3.2 Instrumentations and Apparatus

Filter Paper and Funnels: For removing solid particles from the extract.

Rotary Evaporator: For concentrating the nicotine extract by removing solvent.

Beakers and Flasks: For holding and mixing solutions.

Stirring Rods: For mixing solutions.

Separating Funnels: For liquid-liquid extraction in the acid-base method.

Distillation Apparatus (Round-bottom flask, Condenser, Collection flask): for purification.

Heating Mantle: For heating during distillation.

Safety Gloves: For protecting hands.

3.3 Methodology

3.3.1 Preparation of tobacco powder:

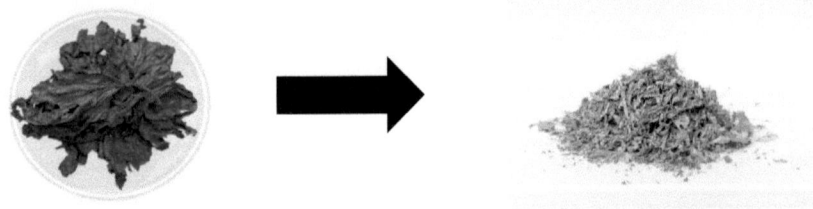

Fig 9 (a): Tobacco Leaves [10]   **Fig 9 (b): Tobacco Waste [10]**

The tobacco waste was dried at 55 °C for 2 hours in a controlled oven to eliminate moisture, preventing degradation and enhancing extraction efficiency. Dried leaves were ground into a fine powder, increasing surface area for optimal solvent interaction. This procedure outlines a rudimentary, potentially hazardous, and inefficient method for attempting to extract and identify nicotine from tobacco waste. It involves a series of chemical processes, including alkaline extraction, solvent separation, steam distillation, and picric acid precipitation. First, 20 grams of powdered tobacco waste are prepared. This increases the surface area, facilitating the extraction of nicotine.

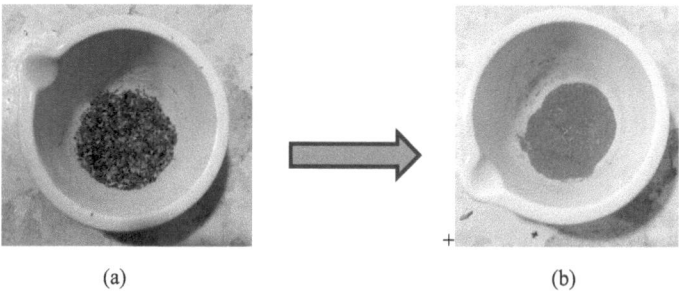

(a)	(b)

Fig 10 (a): Grinded Tobacco Waste (b): Tobacco Powder [Author's own work]

3.3.2. Preparation of Sodium Hydroxide Solution (NaOH)

In a separate beaker, 200 ml of distilled water is used to dissolve sodium hydroxide pellets, creating a basic solution. Sodium hydroxide serves to deprotonate nicotine, converting it into its free base form, which is more soluble in organic solvents. The tobacco powder is then added to the sodium hydroxide solution, and the mixture is stirred for 20 minutes. This allows the nicotine to be released from the tobacco matrix into the aqueous alkaline solution.

Fig11(a): NaOH Solution   **Fig11(b): Tobacco Added in NaOH Solution**
[Author's own work]   **[Author's own work]**

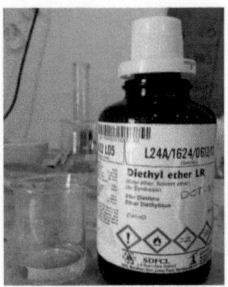 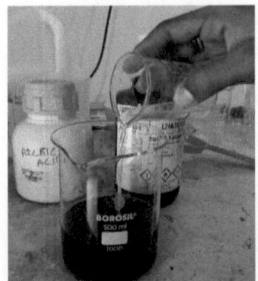

Fig 11(c) Diethyl Ether (DE)   Fig (d): DE Added To Alkaline Solution

[Author's own work]

Next, 25 ml of diethyl ether is added to the mixture. Diethyl ether is a non-polar solvent that preferentially dissolves the free base nicotine. The mixture is transferred to a separatory funnel, where the two immiscible layers—the aqueous (inorganic) layer and the ether (organic) layer—are allowed to separate. Nicotine, being more soluble in ether, migrates to the upper organic layer, while the water-soluble impurities remain in the lower aqueous layer.

The organic layer, containing the extracted nicotine, is carefully separated and collected. This solution is then subjected to steam distillation. Steam distillation is used to vaporize the volatile nicotine along with the ether. Because ether has a low boiling point, it evaporates readily, leaving behind a more concentrated nicotine solution. The resulting distillate is collected in a separate beaker.

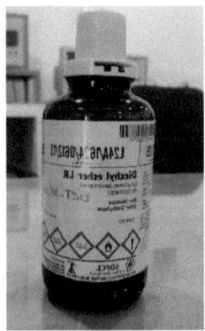 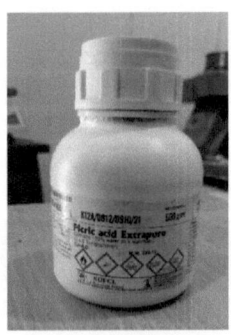 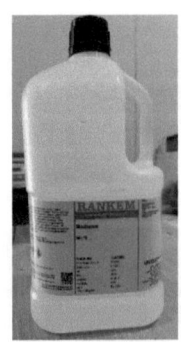

Fig 12(a): Diethyl Ether   Fig 12(b): Picric Acid   Fig 12(c): Methanol

[Author's own work]

To identify the presence of nicotine, 5 grams of picric acid and 10 ml of methanol are added to the distillate. Picric acid reacts with nicotine to form a picrate salt, which is often a crystalline precipitate. The addition of methanol aids in the solubility and crystallization process. The mixture is then placed in a cool water bath for 30 minutes. This cooling process promotes the formation of the picrate precipitate. If a colorless precipitate forms, it suggests the presence of nicotine picrate. However, it is crucial to note several critical points. This method is highly simplified and does not guarantee a pure nicotine product. The resulting precipitate, if any, may contain other extracted compounds from the tobacco. The use of diethyl ether is particularly hazardous due to its extreme flammability and potential for forming explosive peroxides. Furthermore, the handling of sodium hydroxide requires care as it is corrosive. Picric acid is also a hazardous material, being explosive when dry or subjected to shock or heat. This procedure lacks proper purification steps, such as chromatography, which are essential for obtaining pure nicotine.

The statement that "nicotine acid is also known as niacin (Vitamin B3)" is incorrect. Nicotine and niacin are distinct compounds. Nicotine is a toxic alkaloid, while niacin is an essential vitamin. The procedure outlined does not produce niacin. The procedure is designed to isolate nicotine, not produce niacin. The chemical reaction between nicotine and picric acid creates a nicotine picrate salt. In conclusion, while this procedure attempts to isolate and identify nicotine from tobacco waste, it is a dangerous and imprecise method. It should not be attempted without proper laboratory equipment, safety precautions, and a thorough understanding of the involved chemical reactions. The confusion between nicotine acid and niacin highlights the importance of accurate chemical information.

3.3.3 Nicotine Extraction from Tobacco:

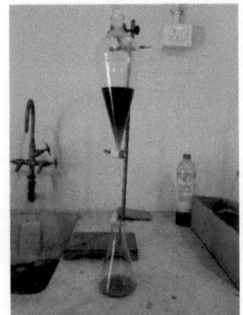  

Fig 13: Separation of the organic phase (Nicotine) from the inorganic phase using a separating funnel [Author's own work]

Nicotine, a potent alkaloid found in tobacco leaves, has garnered significant interest for its pharmacological effects and potential applications. Its extraction, however, necessitates careful consideration of the method employed, as each technique yields varying degrees of purity and efficiency. The experimental design presented utilizes three distinct extraction methodologies: water maceration, ethanol maceration, and acid-base extraction, each with its unique advantages and limitations.

3.3.4 Water Maceration: A Simple Yet Impure Approach

Water maceration, the most straightforward of the three methods, involves soaking powdered tobacco leaves in distilled water for an extended period. The rationale behind this approach lies in the water-solubility of nicotine and other polar compounds. Specifically, 50 grams of powdered tobacco leaves were macerated in 250 mL of distilled water at a controlled temperature of 100 °C, repeated thrice to maximize extraction. This prolonged contact allows for the diffusion of nicotine from the plant matrix into the aqueous solvent. However, the simplicity of water maceration comes at the cost of purity. Water is a versatile solvent, extracting not only nicotine but also a plethora of other water-soluble compounds, including sugars, organic acids, and various plant pigments. This results in a crude extract that necessitates further purification. The subsequent filtration step aims to remove particulate matter, while rotary evaporation concentrates the extract by removing the water, leaving behind a complex mixture. Despite its simplicity, water maceration serves as a baseline for comparison, highlighting the benefits of more selective extraction techniques.

3.3.5 Ethanol Maceration: Enhanced Solubility, Similar Challenges

Ethanol maceration, mirroring the water maceration protocol, substitutes water with ethanol as the solvent. Ethanol, a polar protic solvent, possesses a wider range of solubility compared to water, capable of extracting both polar and moderately non-polar compounds. This characteristic potentially leads to a higher yield of nicotine, as it can solubilize a broader spectrum of compounds present in the tobacco leaves. The procedure involves macerating 50 grams of powdered tobacco leaves in 750 mL of ethanol for 48 hours, repeated thrice, followed by filtration and rotary evaporation. While ethanol's enhanced solubility is advantageous, it also leads to the co-extraction of numerous other compounds, similar to water maceration. These impurities necessitate further purification to isolate nicotine effectively [28].

3.3.6 Acid-Base Extraction: Selective Isolation of Nicotine

The acid-based extraction method stands out for its selectivity, leveraging the basic properties of nicotine to achieve a higher degree of purity. Nicotine, being a weak base, can be protonated in acidic conditions, becoming water-soluble, and subsequently deprotonated in alkaline conditions, becoming soluble in organic solvents. The protocol begins with boiling 50 grams of powdered tobacco leaves in 750 mL of distilled water at 80 ± 5 °C for 20 minutes, enhancing the release of nicotine from the plant matrix. Sodium carbonate (Na_2CO_3) is then added, and the mixture is heated for 10 minutes, creating a basic environment. The resulting mixture is filtered to remove solid particles. The pH of the filtrate is adjusted to 12 using sodium hydroxide (NaOH), ensuring that nicotine is in its free base form.

Liquid-liquid extraction, a crucial step in this method, involves partitioning nicotine between an aqueous phase and an organic phase. Chloroform ($CHCl_3$), a non-polar solvent, is used to extract nicotine from the alkaline aqueous solution. Two extractions with 100 mL of chloroform each ensure maximum recovery. The combined chloroform extracts are then concentrated using a rotary evaporator, yielding a relatively pure nicotine extract [29].

3.3.7 Distillation

Distillation hinges on the variance in boiling points among the components of a liquid mixture. Nicotine, like other volatile compounds, transforms into vapor when heated. This vaporization occurs at a specific temperature, distinct from that of other substances within the tobacco extract. This difference in volatility is the key to separation.

The tobacco extract, a concoction of nicotine and various other volatile compounds, undergoes controlled heating. As the temperature escalates, these volatile substances, including nicotine, transition into a gaseous state. This vapor is then channeled into a cooling apparatus, where it reverts to a liquid form, a process known as condensation. By meticulously managing the temperature during this phase, it becomes possible to isolate fractions of the condensate that exhibit a heightened concentration of nicotine. To achieve optimal purity, the distillation process is often repeated, a technique referred to as fractional distillation. Furthermore, to mitigate the risk of thermal degradation to nicotine, vacuum distillation is frequently employed. This specialized method lowers the boiling point of nicotine, enabling its vaporization at reduced temperatures. This distillation process is a cornerstone of industrial nicotine production and is used in the creation of many products.

Figure 14: Simple Distillation [Author's own work]

3.3.8 Nicotine Extraction Using Solvent Extraction Techniques

Nicotine, an oily yellowish-brown liquid, exhibits hygroscopic properties and demonstrates significant solubility in various solvents, including light petroleum, ether, and alcohol. As a base, it is also miscible with water within a temperature range of 60-210°C. To optimize nicotine extraction from tobacco leaves, several solvents have been evaluated, including isooctane, benzene, petroleum ether, and chloroform. A carefully chosen mixture of ether and petroleum ether can significantly enhance the extraction rate and overall nicotine yield. In this study, the alkaline extract obtained from the preceding step was subjected to solvent extraction using diethyl ether. Specifically, 30 mL of diethyl ether was employed in a separation funnel to partition nicotine from the alkaline aqueous phase. This process was repeated twice to maximize nicotine recovery. The combined ether extracts were then collected in a conical flask and dried using approximately one teaspoon of anhydrous potassium carbonate to remove residual water. Subsequently, the dried solution was filtered to remove the drying agent. Finally, the ether was evaporated using a water bath, ensuring that excessive heat was avoided due to the potential for nicotine hydrolysis at elevated temperatures. Alternatively, evaporation could be conducted at room temperature. The resulting extract was a yellowish oily liquid, indicative of enriched nicotine.

4. Results and Discussion

4.1. Nicotine Extraction:

95% ethanol was employed as the extraction solvent, dispensed via a controlled valve to achieve flow rates of 0.5, 1, and 3 mL per minute. The extraction process was monitored, and samples were collected at 20-minute intervals. The volume of each sample collected was then recorded.

The described procedure for nicotine extraction from tobacco waste, while outlining a series of chemical processes, presents a fundamentally flawed and potentially hazardous approach. It begins with the seemingly innocuous steps of drying and grinding tobacco leaves, aiming to eliminate moisture and increase surface area for enhanced solvent interaction. While these steps are generally sound, the lack of precise temperature control during drying could lead to the loss of volatile compounds, including some nicotine, thereby affecting the overall extraction efficiency. The subsequent preparation of a sodium hydroxide solution to deprotonate nicotine salts and convert them to free-base nicotine is chemically valid, but the absence of specified concentrations and the potential for improper handling of this corrosive substance pose significant risks. Inadequate stirring or contact time during this phase would also result in low extraction yields, compromising the process from its early stages. The introduction of diethyl ether as a non-polar solvent marks a significant escalation in the procedure's hazards. While diethyl ether effectively dissolves free-base nicotine, its extreme flammability and tendency to form explosive peroxides render it an extremely dangerous choice. The separation of the organic layer containing nicotine from the aqueous layer requires a level of skill that is often underestimated, leading to potential losses and contamination.

The distribution coefficient of nicotine between ether and the aqueous phase, a crucial factor for extraction efficiency, is not accounted for, and the presence of other non-polar compounds that may dissolve in ether is not addressed, leading to an impure solution.

The subsequent steam distillation step, intended to vaporize nicotine and ether, introduces further complexities and risks. While steam distillation can separate volatile compounds, it requires specialized equipment and careful control. The high volatility of ether poses a continuous fire hazard, and complete removal of ether from the distillate is challenging, leading to contamination. Furthermore, nicotine can degrade at high temperatures, potentially reducing the yield and purity of the extracted product. The resulting distillate, therefore, is not pure nicotine but a mixture of compounds, necessitating further purification. The identification of nicotine through picric acid precipitation, while a classic method, suffers from a lack of specificity. The formation of a picrate precipitate, while indicative of nicotine, is not exclusive to it, as many other alkaloids can form similar precipitates. The procedure's reliance on this method without further confirmation through techniques like gas chromatography-mass spectrometry (GC-MS) or nuclear magnetic resonance (NMR) spectroscopy leads to ambiguous results. The misconception that "nicotine acid is also known as niacin (vitamin B3)" underscores a fundamental lack of chemical accuracy, highlighting the need for precise knowledge in chemical procedures. Moreover, picric acid itself is a hazardous explosive, particularly when dry, requiring extreme caution in handling. The absence of quantitative analysis, such as HPLC with UV detection, prevents accurate determination of the nicotine content. The lack of detailed information on chemical concentrations, temperatures, and reaction times further compromises the reliability of the results. This makes the procedure more of a qualitative exercise, at best, rather than a rigorous scientific method for nicotine extraction.

In essence, this procedure embodies a dangerous and imprecise approach. The reliance on hazardous chemicals, the lack of purification steps, and the ambiguity of the identification method all contribute to its inadequacy. The misconception regarding nicotine and niacin further emphasizes the necessity of sound chemical knowledge. Any attempt to replicate this procedure without proper laboratory equipment, trained personnel, and reliable analytical techniques is strongly discouraged. A safe and accurate extraction of nicotine necessitates a well-equipped laboratory and highly accurate testing and separation methods.

4.2. Extracted Nicotine Content Analysis:

The extraction methods yielded extracts with distinct physical properties. Aqueous maceration produced the most viscous extract, exhibiting a solid-to-semisolid consistency. Ethanol maceration resulted in a moderately viscous extract, ranging from semisolid to liquid. Conversely, acid-base extraction yielded an oily liquid extract.

The percentage yield and nicotine (NCT) content varied significantly depending on both the extraction method and the leaf position acid-base extraction resulted in the highest NCT content but the lowest overall yield. In contrast, maceration methods produced higher yields, with aqueous extraction yielding more extract than ethanol extraction. However, ethanol maceration resulted in a higher NCT content compared to aqueous maceration, indicating solvent polarity plays a crucial role in NCT extraction.

Fig 15: Extracted Nicotine. [Author's own work]

4.3. Effect of Extraction Method on Nicotine Content

The extraction method significantly impacted the extracted NCT concentration. Acid-base extraction, specific to alkaloid isolation, yielded the highest NCT content despite the lower overall extract yield. This aligns with the principle that acid-base treatment converts non-ionized NCT, soluble in organic solvents, to its ionized salt form, enhancing water solubility. The observed 4.2% yield from acid-base extraction is consistent with previous reports, Maceration, while producing higher yields, resulted in lower NCT concentrations.

4.4. Effect of Leaf Position on Nicotine Content

High-performance liquid chromatography (HPLC) analysis revealed a clear correlation between leaf position and NCT content. The top leaves consistently exhibited the highest NCT concentration, followed selves from the bottom to the top of the tobacco plant. The higher NCT content in top leaves contributes to their desirable aroma and taste, making them preferred in cigarette manufacturing. The observed differences in NCT content across leaf positions, despite similar qualitative chemical composition, highlight the quantitative variations that influence the overall quality and flavor profile of tobacco products.

5. Conclusion

The endeavor to extract nicotine from tobacco waste, utilizing a multi-step chemical protocol comprising alkaline extraction, solvent partitioning, steam distillation, and subsequent precipitation with picric acid, culminated in the acquisition of a product that, although it may contain traces of nicotine, remains substantially impure. Moreover, the methodological framework employed in this process is inherently laden with significant safety risks, stemming from both the chemical reagents involved and the procedural conditions required. The relative simplicity of the extraction sequence, while seemingly advantageous, is undermined by the conspicuous absence of comprehensive purification measures, thereby rendering the overall approach inadequate for the reliable production of high-purity nicotine. This outcome underscores the critical need for more sophisticated and controlled methodologies to achieve both safety and the desired chemical purity in nicotine isolation.

References

[1] Wedad H. Al-Dahhan, Mohammed Kadhom, Emad Yousif, Salam A. Mohammed, Ayad Alkaim, Extraction and determination of nicotine in tobacco from selected local cigarettes brands in Iraq. Lett Appl Nanobioscience. 2021;11(1):3278–90. http://dx.doi.org/10.33263/lianbs111.32783290

[2] Fangyuan Zheng, Qishan Xie, Qingguang Ren, Jilie Kong, Extraction and Purification of Nicotine from Tobacco Rhizomes by Supercritical CO_2, Molecules 2024, 29(5), 1147. https://doi.org/10.3390/molecules29051147

[3] Chen, M., Qin, Y., Wang, S., Liu, S., Zhao, G., Lu, H., Cui, H., Cai, J., Wang, X., Yan, Q., Hua, C., Xie, F., & Wan, L. (2022). Electromembrane extraction of nicotine in inhaled aerosols from tobacco cigarettes, electronic cigarettes, and heated tobacco products. Journal of Chromatography B: Analytical Technologies in the Biomedical and Life Sciences, 1208, 1-10. https://doi.org/10.1016/j.jchromb.2022.122408

[4] Yogesh Jain, Pankaj Bhardwaj Nitin Joshi, India's environmental burden of tobacco use and its policy implications, 20100329January 2024. https://doi.org/10.1016/j.lansea.2023.100329

[5] Saibo Yu, Bingjie Qiu, Yong Jin, Yu Zhao, Wei Luo, Xinhua Qi, Efficient removal of lignin in tobacco stems with choline chloride-based deep eutectic solvents, Volume 226, April 2025, 120634. https://doi.org/10.1016/j.indcrop.2025.120634

[6] Lay-Keow Ng, Michel Hupe, Effects of moisture content in cigar tobacco on nicotine extraction, 1011, Issues1–2, 5 September 2003, Page213-219. https://doi.org/10.1016/S0021-9673(03)01178-6

[7] Hellinghausen G, Lee JT, Weatherly CA, Lopez DA, Armstrong DW. Evaluation of nicotine in tobacco-free-nicotine commercial products. Drug Test Anal [Internet]. 2017;9(6):944–8. http://dx.doi.org/10.1002/dta.2145.

[8] Zhu X-Q, Chen Y, Jia M, Dai H-J, Zhou Y-B, Yang H-W, et al. Managing tobacco black shank disease using biochar: direct toxicity and indirect ecological mechanisms. Microbiol Spectr 2024;12(10) e0014924: http://dx.doi.org/10.1128/spectrum.00149-24

[9] Qayyum I, Fazal-Ur-Rehman, M., & Ibrahim, M. S. (2018). Extraction of nicotine (3- (1-methyl-2-pyrrolidinyl) pyridine) from tobacco leaves separated from gold live classic

brand cigarettes by solvent extraction approach and characterization via IR analysis. Biosciences, Biotechnology Research Asia, 15(4), 799–804. https://doi.org/10.13005/bbra/2688

[10] Jokic S, Gagic T, Knez Z, Benzoic M, & Skerget M. (2019). Separation of active compounds from tobacco waste using subcritical water extraction. The Journal of Supercritical Fluids, 153(104593), 104593. https://doi.org/10.1016/j.supflu.2019.104593.

[11] Zhong, W., Zhu, C., Shu, M., Sun, K., Zhao, L., Wang, C., Ye, Z., & Chen, J. (2010). Degradation of nicotine in tobacco waste extract by newly isolated Pseudomonas sp. ZUTSKD. Bioresource Technology, 101(18), 6935–6941. https://doi.org/10.1016/j.biortech.2010.03.142.

[12] Gong, X.-W., Yang, J.-K., Duan, Y.-Q., Dong, J.-Y., Zhe, W., Wang, L., Li, Q.-H., & Zhang, K.-Q. (2009). Isolation and characterization of Rhodococcus sp. Y22 and its potential application to tobacco processing. Research in Microbiology, 160(3), 200–204. https://doi.org/10.1016/j.resmic.2009.02.004

[13] Patel V.F., Liu F., Brown M. Advances in oral transmucosal drug delivery. J. Control. Release. 2011; 153:106–116. doi: 10.1016/j.jconrel.2011.01.027.

[14] Cilurzo F., Cupone I.E., Minghetti P., Selmin F., Montanari L. Fast dissolving films made of maltodextrins. Eur. J. Pharm. Biopharm. 2008;70:895–900. doi: 10.1016/j.ejpb.2008.06.032.

[15] Matharu, A. S., de Melo, E. M., & Houghton, J. A. (2016). Opportunity for high value-added chemicals from food supply chain wastes. Bioresource Technology, 215, 123–130. https://doi.org/10.1016/j.biortech.2016.03.039

[16] Valverde, J. L., Curbelo, C., Mayo, O., & Molina, C. B. (2000). Pyrolysis kinetics of tobacco dust. Chemical Engineering Research & Design: Transactions of the Institution of Chemical Engineers, 78(6), 921–924. https://doi.org/10.1205/026387600527996

[17] Box G.E., Wilson K. On the experimental attainment of optimum conditions. J. R. Stat. Society Ser. B Method. 1951;13:1–45. Doi : 10.1111/j.2517-6161.1951.tb00067.x

[18] Tita, G. J., Navarrete, A., Martín, Á., & Cocero, M. J. (2021). Model assisted supercritical fluid extraction and fractionation of added-value products from tobacco scrap. The Journal of Supercritical Fluids, 167(105046), 105046.

https://doi.org/10.1016/j.supflu.2020.105046 .

[19] Dai, J., Kim, K.-H., Szulejko, J. E., & Jo, S.-H. (2017). A simple method for the parallel quantification of nicotine and major solvent components in electronic cigarette liquids and vaped aerosols. Microchemical Journal, Devoted to the Application of Microtechniques in All Branches of Science, 133, 237–245. https://doi.org/10.1016/j.microc.2017.02.029

[20] Kheawfu, K., Kaewpinta, A., Chanmahasathien, W., Rachtanapun, P., & Jantrawut, P. (2021). Extraction of nicotine from tobacco leaves and development of fast dissolving nicotine extract film. Membranes, 11(6), 403. https://doi.org/10.3390/membranes11060403

[21] Ranjana Singh, Extraction and Estimation of Nicotine Present in Different Tobacco products, 2021 JETIR October 2021, Volume 8, Issue 10.

[22] Pal, R. Megharaj, M. Kirkbride, K. P. & Naidu, R. (2013). Illicit drugs and the environment--a review. The Science of the Total Environment, 463–464, 1079–1092. https://doi.org/10.1016/j.scitotenv.2012.05.086

[23] Ferguson S.G, Nichols D, Patel R., Jacobson G.A. Determination of Nicotine in Cartridge-Based Electronic Cigarettes. Anal. Lett. 2015; 48:2715–2722. Doi: 10.1080/00032719.2015.1048349.

[24] Anyiam Ngozi Donald, Extraction and Estimation of the Amount of Nicotine in a Tobacco Leaf, s2020 IJCRT | Volume 8, Issue 8 August 2020 | ISSN: 2320-2882

[25] Mulyadi AF, Wijana S, Wahyudi AS. Optimization of nicotine extraction in tobacco leaf (Nicotiana tabacum L. Study: Comparison of ether and petroleum ether) 2013.

[26] Almeida R.N., Rossa G.E., de Castro J.H., Cavassa A.F., Vargas R.M.F., Cassel E. Extraction and fractionation of long pepper essential oil: Process modeling and simulation. Braz. J. Chem. Eng. 2023; 40:1103–1113. Doi :10.1007/s43153-023-00307-0.

[27] Naviri Roja 1, Dr. B. Mahesh Babu2, EXTRACTION OF NICOTINE FROM TOBACCO, 2025, IRJET | Impact Factor value: 8.315 | ISO 9001:2008 Certified Journal | Page 703

[28] Corkery JM, Button J, Vento AE, Schifano F. Two UK suicides using nicotine extracted from tobacco employing instructions available on the Internet. Forensic Sci Int [Internet].

2010. http://dx.doi.org/10.1016/j.forsciint.2010.02.004

[29]   Baker R. Smoke generation inside a burning cigarette: Modifying combustion to develop cigarettes that may be less hazardous to health. Prog Energy Combust Sci. 2006;32(4):373–85. http://dx.doi.org/10.1016/j.pecs.2006.01.001

YOUR KNOWLEDGE HAS VALUE